POLAND

Major World Nations

POLAND

Julian Popescu

CHELSEA HOUSE PUBLISHERS

Philadelphia

Chelsea House Publishers

Copyright © 2000 by Chelsea House Publishers,
a division of Main Line Book Co.
All rights reserved.
Printed in Malaysia

First Printing.

1 3 5 7 9 8 6 4 2

Library of Congress Cataloging-in-Publication Data

Popescu, Julian.
Poland / Julian Popescu
p. cm. — (Major world nations)
Includes index.
Summary: An overview of the history, geography, economy,
government, people, and culture of Poland.
ISBN 0-7910-5394-6
1. Poland Juvenile literature. [1. Poland.] I. Title. II. Series.
DK4042.P67   1999
943.8'05—dc21   99-13405
CIP

ACKNOWLEDGEMENTS

The Author and Publishers are grateful to the following organizations and individuals
for permission to reproduce photographic illustrations in this book:
Polish Tourist Information Center; Father Andrew Janicki, and Mrs. Irena Bielatowicz

CONTENTS

POLAND

0   100   200   300   Km.

0   100   200 Miles

BALTIC

SEA

RUSSI

Gulf of Pomerania

Gulf of Gdansk

LITHUANIA

GDYNIA
SOPOT
GDANSK

Lagoon

Masurian Lakes

Pomeranian Lakes

SZCZECIN

R. Odra

R. Notec

TORUN

R. Vistula

PLOCK

R. Bug

GERMANY

WLOCLAWEK

Lake Goplo

WARSAW

POZNAN

R. Warta

R. Odra

SILESIA

LITTLE

LODZ

R. Pilica

LUBLIN

R. Wieprz

POLAND

WROCLAW

SUDETES   Mt.Sniezka △

CZESTOCHOWA

R. Vistula

Lublin
Highlands

R. Nysa   NYSA

SANDOMIERZ

LW

R. San

BESKIDS

KATOWICE

NOWA HUTA

CARPATHIAN MTS.

KRAKOW

R. Dunajec

CZECH

REPUBLIC

ZAKOPANE

TATRAS

△ Gerlachowka
Peak

SLOVAKIA

FACTS AT A GLANCE

Land and People

Official Name Republic of Poland

Location Lies at the center of the European continent

Area 121,000 square miles (315,000 square kilometers)

Climate Temperate

Capital Warsaw

Other Cities Krakow, Katowice, Lublin, Gdansk, Poznan, Wroclaw

Population 39 million

Major Rivers Vistula, Odra, Dunajec, Wieprz, Bug, Dnepr

Major Lakes Masurian, Pomeranian

Mountains Carpathian, Beskids

Highest Point Rysy Peak (8,200 feet/2,499 meters)

Official Language Polish

Ethnic Groups Polish, 97.6 percent; German, 1.3 percent; Ukrainian, .6 percent

7

Religions	Roman Catholic (95 percent)
Literacy Rate	99 percent
Average Life Expectancy	68.6 years (male); 77.16 years (female)

Economy

Natural Resources	Coal, lead, natural gas, copper, sulfur
Division of Labor Force	Industry, 29.9 percent; agriculture, 26 percent; services, 44.1 percent
Agricultural Products	Wheat, dairy products, potatoes, poultry, beef
Industries	Iron and steel; coal mining, textiles, machinery, chemicals
Major Imports	Food, fuel, machinery
Major Exports	Machinery and transportation equipment, chemicals, consumer products, food, fuel
Major Trading Partners	Germany, Italy, Russia
Currency	Zlaty

Government

Form of Government	Unitary multiparty republic
Government Bodies	Senate, Sejm, Council of Ministers
Formal Head of State	President
Head of Government	Prime minister
Other Chief Officials	Deputy prime minister
Voting Rights	All citizens 18 years of age and older

HISTORY AT A GLANCE

966 A.D. King Mieszko I brings Christianity to the Polish people and Poland is recognized as a separate principality by Rome. He is the first king of the Piast dynasty that would rule Poland until the 14th century.

967–990 Mieszko conquers much of the land surrounding his principality and Poland becomes one of the strongest countries in Eastern Europe.

992-1025 Boleslaw I succeeds his father and continues to increase Poland's landholdings.

1058 King Boleslas II becomes king and makes Krakow the capital city of Poland.

1079 During an uprising against King Boleslas II, Stanislaw, the bishop of Krakow, is killed by order of the king. Stanislaw later is canonized and becomes a symbol of resistance to illegal state authority.

11th–13th centuries The Catholic Church is the center of learning and art in Poland. Various religious orders come to Poland, among them the Benedictines, the Cistercians, and the Dominicans.

1241	The Battle of Lignica takes place with Henry the Pious leading the Poles against the Tartar invasion. Henry and most of his knights do not survive but the Tartars are forced to flee the country.
1386	Polish Queen Jadwiga marries the Duke of Lithuania uniting their two countries.
1410	The Teutonic knights of Germany, who had been raiding Poland and Lithuania, are defeated at the Battle of Grünwald.
1473	Nikolay Kopernik (Copernicus) is born in Torun.
early 15th century	Copernicus studies astrology and writes a book making the revolutionary claims that the earth rotates on its axis and orbits around the sun.
1572	After the death of King Zygment August, the Polish knights begin to elect their kings. At the end of the 16th century Poland is the largest state in Europe.
1587–1632	King Sigismund III reigns and moves the capital city to Warsaw.
16th and 17th centuries	During what is called the golden age of Polish history, Poland expands its borders, builds beautiful castles and cathedrals, and consolidates its holdings in eastern Europe.
1683	The Polish hussars (cavalry) under King Sobieski go to the rescue of the city of Vienna, under attack by the Turks.
1772	Poland is partitioned by its powerful neighbors. Russia takes the eastern provinces, the Austrian Empire is given the southern area, and Prussia assumes control of Pomerania.

1794	Polish General Thaddeus Kosciuszko leads an insurrection against the Russians and defeats them at the Battle of Raclawice.
1795	Poland again loses her freedom to Russia and is partitioned once more. Poland's future national anthem is written by Josef Wybicki while he is in Italy with Polish troops, trying to build an army to regain Poland's freedom.
18th century	The city of Warsaw flourishes with beautiful palaces being built and industry developing on a huge scale.
1806	Napoleon helps the Polish army reclaim their country.
1815	After Napoleon's defeat at Waterloo, Poland is again divided with a large part of its land going to Russia.
1863	What is called the "January Insurrection" takes place with a call to the Polish people to rise up and declare their freedom.
1914–1918	In World War I, Poland joins forces with Germany and the Axis against Russia and the Western powers. After the war, the Austro-Hungarian Empire breaks up, and Poland once again becomes an independent country.
1939	Adolph Hitler and his German army invade Poland, beginning World War II.
1939–1945	The Germans set up concentration camps in Poland where millions lose their lives. The Polish people, with the help of the Russian army, finally drive the Germans out. Before they leave, the Germans devastate the country.

1945	The Soviet Union imposes a Communist system of government on Poland.
1950s and 1960s	The cities of Poland rebuild from the ashes of World War II. The Communist party suppresses freedom of speech, takes over most of Poland's lands and industry for the state. The Catholic Church manages to keep some independence.
1980s	In August the first independent trade union in Eastern Europe is formed in Poland called Solidarity, after protests and strikes led by Lech Walesa. Its membership quickly swells to 10 million people. Solidarity becomes a force in Poland and a threat to Poland's Communist government.
1981	The Communist government declares martial law to try to destroy the Solidarity movement.
1989	Poland holds its first free election since before World War II. Lech Walesa, the head of Solidarity is elected president and parlimentary elections remove most of the Communist Party from their seats. The Polish example inspires other countries of the Soviet bloc and by the end of the year Communism has collapsed.
1995	Walesa loses the presidential election to Alexsander Kwasniewski.
1997	In May a new constitution that took eight years to complete is adopted. Poland's economy continues to improve with inflation down.

1

Between East and West

Poland is part of the Great European Plain which stretches across Central Europe from the Low Countries in the west, across the whole of Germany and far into the Ukraine and Russia in the east. Poland means "land of fields." This gently undulating plain is broken up by occasional ridges, by woods and marshes and important waterways such as the Odra (Oder) and Vistula rivers. In the south there are hills and valleys and snowcapped mountains.

In the winter the bitter winds blowing from the Arctic and Siberia freeze the rivers, lakes, and marshes. These winds bring heavy falls of snow which cover the land from end to end. But in

Beautiful scenery and snow-capped peaks in southern Poland. The view from Zakopane.

spring the country comes to life. The fields take on a fresh green color and masses of flowers border the roads. The orchards close to villages are like clouds of blossom on the distant horizon.

Today Poland is a Republic. With an area of 121,000 square miles (315,000 square kilometers) the country is over twice the size of England and Wales taken together. Its population of over 39 million is made up mostly of Poles with a few national minorities such as Ukrainians, Ruthenians, Lithuanians, Russians and Tartars in the east and Germans and Czechs in the west and south. Poland occupies seventh place in Europe in size of population coming after Russia, Germany, Great Britain, Italy, France, and Spain.

The Poles are a generous and hard-working people. They belong to the Slav family of peoples and are related to the Russians, Ukrainians, Czechs, Slovaks, Slavs, and Bulgarians. Because of their troubled history and sufferings at the hands of foreign invaders, the Poles are very fond of their country and proud of their nation. Though many Poles live in scattered communities all over the world, they keep in touch with their relatives at home, and observe all their customs and festivals.

A little more than half a century ago the shape of Poland was very different from its shape today. The country had a short coastline and its borders reached as far south as Romania. But the borders were changed after the Second World War. A large part of eastern Poland became Russian territory and, in exchange, Poland was given part of what used to be called East Prussia and Eastern Germany.

In 1944 the communists came to power in Poland with the help of the Russians, the country was ruled by one party. That party was called the Polish United Workers Party and it was officially founded in December 1948. The provisional Polish Government also signed a Treaty of Friendship and Mutual Aid with the Soviet Union on April 21, 1945 which bound the country closely to Soviet policy on foreign affairs. Then, on June 14, 1955, a conference of communist countries was held in Warsaw. It decided to set up the Warsaw Pact Organization, an alliance of communist countries opposed to the North Atlantic Treaty Organization (NATO) to which Great Britain, the United States and other western countries belong. After the fall of communism in 1989, Poland would petition to become a member of NATO.

During the first half of this century Poland was a mainly agricultural country and the majority of Poles were peasants. They lived in whitewashed and thatched cottages and owned small farms. They cultivated rye, wheat and potatoes and used horses in simple harness to draw wagons and plows. Some of the peasants worked as laborers for big landowners who lived in manor houses and cultivated sugar beet, hops, and tobacco which they sold for cash. They also reared the famous Polish horses. The peasants' daughters worked as servants for the rich landowners, or gentry as they were called. Young children were often sent to keep an eye on the cattle grazing the open fields or the flocks of geese on the village common.

After the rise of communism, life in Poland changed when the big landowners lost their property to the state or emigrated abroad. Most of the land was then farmed by the peasants, though

These dancers are wearing traditional costumes—as worn by Polish peasants for many centuries.

a few new state farms were also set up. Many factories and powerstations were built all over the country and Poland became an industrial nation. Peasants and their families left the land and went to the growing towns to become factory workers and earn good wages.

The unit of money Poles use is called zloty (pronounced *zuoty*). Each zloty consists of 100 groszy (pronounced *groshy*). The zloty is therefore a decimal currency. The groszy is almost worthless and three zloty will buy a cup of coffee or a bottle of beer.

The towns, too, changed from year to year. New housing estates and apartments are built. New streets, shopping precincts, government buildings and squares appear on the edges of towns. Everywhere the visitor can see engineering factories and textile

16

mills, oil refineries and chemical plants belching fumes and white sulphurous smoke.

The Polish countryside has also changed in the last 50 years. Roads have been widened and signposting improved for the growing traffic. Rivers have been dammed for irrigation and to produce hydroelectric power. High-tension cables cover parts of the country like giant cobwebs. Some of these cables are connected to those of neighboring countries, such as the Czech Republic, Slovakia, and Russia, forming a grid to even out peak demands for electricity in very cold weather. Oil and gas pipelines have also been built to speed the transport of these precious and expensive fuels.

Poland is a Central European country which shares in the traditions and way of life of both east and west. Germany is Poland's most important trading partner and Polish industry depends on substantial imports of raw materials.

Unfortunately for the Poles, their country has long open borders and is sandwiched between two powerful neighbors, Russia in the east and Germany in the west. On many occasions during their long history, the Poles had to fight long wars with both their neighbors to stay free and independent. They have lived through terrible times of destruction and famine. This has made them distrust both Russians and Germans.

Nevertheless, Poland is an hospitable country whose people enjoy the good things of life. Castles, palaces and cathedrals built by Polish master masons with the help of German and Italian architects in the Gothic and Renaissance style can be seen in many cities. Some of the castles are medieval forts with drawbridges, stout ramparts and narrow windows. The royal castle on Wawel Hill in Krakow was burned down and then rebuilt as a Renaissance palace by Italian architects in the 16th century. Now it has been turned into a museum for people to visit and enjoy.

Poland also has many splendid lakes for yachting and mountain slopes for skiing. Let us visit this country in the heart of Europe and look more closely at the land and at the way of life of the Polish people.

Sailing on one of the Masurian lakes in eastern Poland.

2

Lowlands and Mountains

Looking at the map, we can see that Poland today is almost square in shape. It has a long stretch of the Baltic coast from the Gulf of Gdansk in the east to the Gulf of Pomerania in the west, into which the Odra River flows. The Baltic Sea is shallow in many parts with a depth of only about 120 feet (36 meters) and there are sandbanks and shoals along the Polish coast which make navigation difficult. But the Baltic Sea is Poland's most important outlet and waterway.

To the east and northeast, Poland has a land border with Russia, Lithuania, Belarus, and the Ukraine. The border runs from the Carpathian Mountains in the south, across a flat plain, along part of the Bug River and then across the plains again to the Gulf of Gdansk. Germany is Poland's neighbor in the west from whom it is separated by the Nysa and Odra rivers. To the south is the Czech Republic and Slovakia. This is an old border, running along the crests and valleys of the Carpathian Mountains and their foothills.

19

A sandy beach on the Baltic coast.

Thousands of years ago, thick ice and snow covered northern Europe rather as Greenland or the Arctic Sea are covered today. That period in the Earth's history is known as the Ice Age. The ice sheet had advanced southward from Scandinavia, across the Baltic Sea and into the lowlands of present-day Poland churning the soil as it went along. When the climate became warmer and the ice sheet melted away, ridges of rock and earth were left behind. These ridges are called moraines and they can be seen today as rounded hills, steep escarpments or heaps of boulders. In between the moraines shallow lakes and bogs were formed. The Ice Age left a big boulder called Devil's Stone in Pomerania measuring 82 feet (25 meters) across.

The north of Poland consists of the Baltic coast regions and a number of small islands in the Baltic Sea where gray seals live. The coast is a land of sand dunes which the winds shift from place to place. The sand dunes are made of sharp and gritty sand. They have gentle slopes facing the wind and steep slopes away from the wind. Clumps of coarse grass and sedges grow among the sand dunes. Saltwater lagoons and lakes abound here. Little or no farming can be carried out because the soil is so poor.

Moving south we come to the Baltic lake zone where we find glacial lakes, marshes and morainic hills dating from the Ice Age about 115,000 years ago. The Pomeranian Lakes are in the west and the Masurian Lakes in the east. This is an area of pine trees with red bark. They thrive on the hills while the lowlands provide peat for heating. Drainage is so poor that the land becomes waterlogged after heavy rain. Reeds, rushes and clumps of rye grass are the main vegetation. Movement along the marshlands is possible with flat-bottomed boats. In some parts the marshland has been drained and turned into good farming land. Swans, cormorants, and cranes live in the marshes. In summer, thick swarms of mosquitoes fill the air.

Most of Poland is lowland and meadows which are well watered with rivers and small lakes where carp are bred on a large scale. The soil of these lowlands is sandy and needs careful cultivation with the help of fertilizers for good crops of wheat, rye and potatoes to be grown. The meadows provide hay for cattle and horses.

The Vistula River (*Wisla* in Polish) is Poland's main waterway and divides the country into two halves. When swollen by heavy rain, the river carries down much sand and silt and deposits them in its lower reaches and in the estuary linked to the Gulf of Gdansk. The old capital of Krakow and the new capital of Warsaw are situated on the banks of the Vistula.

The Old Highlands of the upper Odra and Vistula rivers consist of fertile land reaching into the foothills of the Carpathian Mountains. The eastern part of the country is called Silesia, the middle section is Little Poland and the eastern part is the Lublin Highlands. The land yields rye, wheat and sugar beet and there are many fruit orchards and beehives in the valleys.

Silesia lies near the border with the Czech Republic and Germany and is in the upper basin of the Odra River which rises in the Carpathian Mountains. Silesia is also called the Black Country because of its many coal mines and iron and steel foundries. Near the city of Katowice there are large deposits of anthracite. Coal suitable for coking is also mined and exported across the border to the factories of the Czech Republic and Slovakia. Silesia also has iron quarries, zinc mines, and rock salt mines.

Western Silesia is hilly country with dense forests of spruce, pine, fir, beech and ash. Woodcutters work long hours in these forests using their axes or mechanical saws to cut trees for the sawmills and furniture factories in the valleys. Winter holiday resorts have also been built because the country has fine ski slopes. Eastern Silesia is fertile lowland where maize, wheat and vegetable crops are grown to feed the workers in the nearby industrial cities.

Let us look at the map again. Little Poland lies right in the heart of the country. It has well-drained rich earth with a good rainfall. Wheat and sugar beets are grown there and the farmers rear many

The royal castle at Lublin.

pigs from which the famous Polish ham and bacon are made. There are also important forests of oak, some 400 years old, and beech. Iron ore deposits are mined near Częstochowa. The surrounding countryside has many towns and villages with beautiful churches and castles.

The Lublin Highlands or Plateau takes its name from the important market town of Lublin which lies to the north where several trade routes meet. Lublin has a royal castle, sugar refineries and flour mills. The Highlands range from 656 to 1,640 feet (200 to 500 meters), and consist of chalk with a thick covering of black earth. Pine trees grow on the hills and hazelnut trees thrive in the valleys. Here and there are small farms with whitewashed houses and outbuildings. The peasants grow wheat and sugar beet.

A mountain lake amid some of Poland's more rugged scenery

The Carpathian Mountains which form the whole of southern Poland are a continuation of the Alps. In the west they are called the Sudetes Mountains. A mountain range called the Giant Mountains is shared with the Czech Republic. Mount Sniezka, its highest peak, rises 5,259 feet (1,603 meters). Further east are the snowcapped High Tatras whose highest peak, Gerlachovka Peak, rises to 8,737 feet (2,663 meters) and lies just across the border in Slovakia. Poland's highest peak, the Rysy Peak which rises to 8,200 feet (2,499 meters), is found in the Tatras. The mountains are formed of hard granite rock with steep sides and fantastic shapes.

There is a Polish story about a timber cutter called Black Dimitriy whose axe was so sharp he could cut water with it. People said he cut a mountain stream into lengths and stacked them up like logs. Then he gave a party for the seven best axemen of the seven highest peaks of the Tatra Mountains. When they danced a jig, the sparks from their hobnailed boots flew into the sky and stayed there as stars. They stamped the ground with their feet and caused an earthquake. They shouted so loud, the mountain streams were frightened and ran away. If you look at those mountain streams today you can see that they are running as fast as they can. Because of their natural beauty, the Tatras have been turned into a National Park.

In the east there are two important mountain ranges, the West and East Beskids, composed of limestone and crystalline rocks. They have rounded summits which rarely exceed 4,265 feet (1,300 meters) in height. Thick dark-green forests of spruce, larch and Scots fir cover the lower part of the mountain sides. Above them are the high pastures where flocks of sheep and herds of cattle graze when the winter snow has gone. But every autumn before the fierce weather starts, they are driven down to the shelter of the villages.

In winter the Beskids are covered with snow which thaws only early in May. After the snow has gone, brambles, brushwood, ferns, mountain flowers and grass grow in great profusion. Birds of prey, such as the golden eagle, can be seen wheeling across the valleys in search of prey. Wild animals, wild sheep, timber wolves, stags and deer roam the mountain sides. And, not long ago, brown

Today bison are protected animals living in national parks.

bears and wild boar lived here too. Squirrels and their close relations the marmots climb the fir trees. Aurochs and bison roamed southern Poland in days gone by but were killed off by hunters. Today bison, brown bears and elk are protected animals living in national parks such as the Bialowieza Forest famous for its ancient oak trees.

It can be very hot in Poland's lowlands during the summer and very cold in winter, but most of Poland has moderate temperatures. The average temperature in January in the west of the country is 30 degrees Fahrenheit (*minus* one degree Celsius) while the average temperature in July is 66 degrees Fahrenheit (19 degrees Celsius). In central Poland the average temperature is 27 degrees Fahrenheit (*minus* 3 degrees Celsius) in January and 66 degrees Fahrenheit (19 degrees Celsius) in July. In the north of the country westerly winds predominate which makes the climate damper. Rainfall is lowest in the lowlands and highest in the mountains of the south.

3

The Vistula River

The Vistula River is 677 miles (1,092 kilometers) long of which 666 miles (1,072 kilometers) belong to Poland. It is one of the longest rivers to flow through one country in Europe.

The Barania Gora, or Ram's Mountain, is in the West Beskids. Two small streams, the Little White Vistula and the Little Black Vistula rise on the slopes of the Barania Gora rushing down among the rocks and forming small waterfalls. When the two streams join at the foot of the mountain, they suddenly become

Shooting the rapids on the Dunajec River.

more powerful as they continue descending the mountain slopes. The new stream brings down with it broken rocks and boulders which it deposits in heaps lower down the valley. The stream then turns swiftly in a northerly direction and flows along a green valley with mountains towering high above it on either side. Trout live in the clear mountain water. Where trees have been cleared in the valley there are Alpine pastures with dairy cattle, goats, and sheep.

About 28 miles (45 kilometers) from the industrial town of Katowice and the surrounding coal-mining area, the Vistula River has been dammed, at Goczalkowice in Silesia. The water behind the dam forms a reservoir which provides drinking water for the towns of Silesia. The dam also regulates the flow of the water which can swell quickly when heavy rains fall in the mountains. Just beyond Goczalkowice the Vistula is joined by the Premsza River. The river now becomes navigable and coal barges go down stream to the industrial city of Krakow with Nowa Huta, an iron and steel industrial section founded in 1949. *Nowa Huta* means "The New Foundry."

As the Vistula continues its journey, it is joined by the waters of the Dunajec River which has made its way through a deep gorge in the sharp gray-white Pieniny Mountains. Then it is the turn of the San River to run into the Vistula, near the town of Sandomierz, where another concrete dam with sluice gates regulates the flow of the San which brings water from the Carpathian Mountains. In its upper reaches this river forms the border with the Ukraine.

Beyond its junction with the San River, the valley of the Vistula widens considerably, the current slows down and the river mean-

The house in which the composer Chopin was born.

ders forming sandbanks. It is wide but shallow with reeds growing along the banks and into the water. In winter, the whole river freezes for more than two months. When the thaw comes the waters flood the flat countryside as far as the eye can see. Flat ferryboats are used to carry people and vehicles from place to place and from one bank to another.

Before the Vistula reaches the capital city of Warsaw it is joined by two more rivers, the Pilica from the west and the Wieprz from the east. The river is now deep enough to carry small steamers and large barges. One of the bridges spanning the river at Warsaw carries the main railway line from Berlin to Moscow.

North of Warsaw, the Vistula veers westward and passes by a village called Zelazowa Wola where the famous composer Frederik

Chopin was born in 1810. Chopin played the piano and composed a great deal of piano music, much of it based on old Polish dances such as the polka and mazurka.

Just beyond Warsaw, the Vistula is joined by its main tributary the Bug River. Swollen in size by the waters of the Bug, the Vistula widens considerably. There are islands and sandbanks in the middle of the river. Reeds and bulrushes cover the islands and storks and cranes nest there. The land on the river banks is rich and black: wheat and sugar beet are grown in vast fields.

Past the city of Plock, the Vistula becomes a reservoir for the next 31 miles (50 kilometers) because its course has been dammed at Wloclawek and a hydroelectric power station has been built there. The electricity produced by this station is fed into Poland's national grid.

By the time the Vistula has reached the old city of Torun it is

A view across the Vistula from Kazimierz Dolny.

very wide. Road and rail bridges span its waters and soon it is joined by canal to the Notec River and the Warta River. Goods can now travel by tug and barge from the Odra River to the Vistula, via the canal, before going up the Bug River which is linked by canal to the Dnepr River and to the Russian waterway system.

The Vistula now turns north heading for the sea. The river varies in width a good deal in its lower reaches, sometimes being fairly narrow and deep and at others quite wide with terraced sandbanks in the middle. The color of the muddy water is chocolate brown. The banks have many weeping willows and patches of swamp full of mosquitoes, and frogs which call to each other late in the evenings.

Before it reaches the sea, the Vistula branches into two arms to form a delta, a vast area of marshes and islands. The main arm is called the Martwa Wisla, or Dead Vistula. Mud and sand which the river has brought from its upper reaches are deposited here. Just before it empties its waters into the Gulf of Gdansk, the Dead Vistula branches out again running parallel with the coast and enters the sea at Gdansk. The eastern arm empties its waters into a wide lagoon, some 37 miles (60 kilometers) long. The shores of the lagoon are wide and sandy. Fortune hunters search there for amber (fossilized resin) and gum. Amber is valuable and is used for making beads and jewelry. In the Middle Ages, because there were few roads, the Vistula was the most important route between the north and south of Poland. Merchants loaded the precious amber onto boats and carried it up the Vistula to the south to be exchanged for gold, silver and furs.

4

Early History and Kings

Thousands of years ago Poland was a wild country covered by vast forests and undergrowth broken up by marshes and lakes. In the few clearings there were primitive human settlements of Baltic and Lusatian tribes, who lived in log huts and tilled the land to grow crops.

Meanwhile, the Poles belonged to the tribes of Slav peoples wandering across the grassy plains of Central Asia and Russia. They lived in tents and owned horses which helped them to move long distances carrying all their personal belongings. They also kept domestic animals, mainly cattle and dogs.

One of these Slav tribes, the Czechs and Slovaks, moved westward being pushed from behind by the warlike barbarians called Avars who built their settlements in circles surrounded by moats and earth ramparts. During the sixth century A.D. the Czechs and Slovaks settled in the valleys and plains south of the Carpathian Mountains. In the ninth century, the Czech kingdom of Moravia owned much land in the southern part of present-day Poland.

Further north the Polish tribes, again driven westward by marauding barbarians, settled on the banks of the Vistula as far north as the Baltic coast. An old legend says that three brothers Lech, Czech and Rus came to the lands of present-day Poland. Czech went south and founded Czechoslovakia and Rus went east and founded Ruthenia. Lech stayed behind and wandered east of the Vistula until one evening he saw a white eagle on the branch of a tree against the red sky. He decided to settle near the eagle's nest. Since then the emblem of Poland has been a white eagle on a red background.

The Polish tribes were often at war with each other. They needed a leader to unite them, to turn the country into a strong kingdom safe for people to live in, protected from the raids of barbarians. According to an old legend the first prince to unite the Polish tribes was Prince Popiel, but plotters carried him away and locked him in a tower in the middle of Lake Goplo where a multitude of mice devoured him. So it was a humble peasant and wheelwright, called Piast, who was elected king of the people, and founded the Piast dynasty which ruled Poland from the 10th to the 14th centuries.

The first real king to be recorded in the history of the Polish people was Mieszko I. He brought Christianity to his people in 966 A.D. So, at an early stage in their history, the Polish people became Roman Catholics. And Mieszko, the first Christian king of the Polish people, was grandfather of King Canute the Great, ruler of England, whose mother, Sigrida, had married Forkbeard of

Denmark. This established the first direct link between the kingdoms of England and Poland.

Boleslas II, nicknamed the Bold, came to the throne in 1058 and was the first Polish king to reside in Krakow, Poland's medieval capital. The king behaved badly during his reign of 21 years. Following plots among his nobles, he murdered Stanislas, Bishop of Krakow, at the altar of his church just as the King of England had Thomas à Becket put to death.

Another descendant of Piast, Henry the Pious, Duke of Silesia, will always be remembered for saving Poland from the scourge of the Tartar invaders. The word Tartar means "inhabitant of hell." The Tartars had slit eyes and high cheekbones. They were swift and hardy horsemen, armed with bows and arrows, swords, round shields, and spears. They wore fur hats and coarse leather boots trimmed with fur. A big battle was fought at Lignica in 1241 in which Henry and all his knights perished but so many Tartars were killed that those who were left fled back to Asia.

The last king of the Piast dynasty, Casimir the Great, was a wise ruler. He gave his country many laws which were collected in handwritten books for easy reading and, in 1346, he founded the University of Krakow, the second oldest university in Europe. He built many new towns and people later said of him that he found a Poland of wood and left one of stone. Casimir the Great also allowed Jews to settle in Poland at a time when Jews were persecuted and driven out of many European countries.

In 1386 the Polish Queen Jadwiga, grandniece of Casimir the Great, married Jagiello, Duke of Lithuania, thus uniting Poland

A procession of professors taking part in the celebrations for the 600th anniversary of the Jagiellonian University in Krakow.

and Lithuania into a great country. This union was important because German knights raided and seized Polish and Lithuanian lands from time to time. The German knights had formed themselves into a military order called the Teutonic Knights. They wore a uniform which was a white tunic with a black cross on it. After Poland's union with Lithuania, the Teutonic Knights declared war on Poland. A great battle was fought at Grünwald in 1410 in which the Teutonic Knights were defeated and so lost most of their lands on the Baltic coast.

The Piast dynasty was followed by the House of Jagiellons. Zygmunt August, the last Polish king of the Jagiellons died in 1572.

After that, the knights of Poland elected their kings. The election
of the new king used to take place in the village of Wola, near
Warsaw. Knights and nobles from all over the country took part in
the election.

The reign of the Jagiellonian dynasty has been called the Golden
Age of Poland because at the end of the 16th century Poland was
the largest state in Europe, including in its provinces Lithuania,
Ruthenia and Moldavia.

King Sigismund III came to the throne in 1587 and reigned until
1632. He transferred the capital of the country from Krakow to
Warsaw. The statue of King Sigismund III in Warsaw's Castle
Square is the city's oldest monument.

In 1655 the Swedes invaded Poland. They occupied Poznan and
Warsaw and even took Krakow in the south. They laid siege to the

fortified monastery of Czestochowa but failed to capture it. The Swedes finally withdrew. A few years later the Polish cavalry was sent to attack the Swedes in their own country. The Polish troops crossed the sea by Denmark swimming with their horses, and taking the Swedes by surprise, defeated them.

Another interesting king of Poland was John Sobieski III who ruled the country from 1674 to 1696 at a time when the Turks wanted to conquer Europe. In 1683, the powerful Turkish army of infantry—*janissaries*—backed by archers with barbed arrows, and horsemen, called *spahis*, began the siege of Vienna. Months went by and the stocks of food dwindled in the besieged city. The situation became desperate. So the German Emperor Leopold I asked the Polish King Sobieski for help. King Sobieski agreed and mustered his troops. They set off on the long journey across the mountains and through valleys. They finally arrived near Vienna and prepared to attack the Turks whose camps they could see in the distance. Leading the charge of the Polish cavalry called hussars, who were clad in steel armor and wore wings on their shoulders to make them look fierce, King Sobieski crushed the Turkish army just under the walls of Vienna. He entered the city in triumph and was cheered by the population. King Sobieski also captured a magnificent embroidered tent with hundreds of small mirrors, weapons studded with jewels, and carpets belonging to a vizier—a high-ranking Turkish commander. These trophies can be seen on display to this day in Wawel Castle in Krakow.

During the 18th century the borders of Poland stretched from the Baltic Sea almost to the Black Sea. Being the third largest coun-

The Wawel Castle in Krakow, where the trophies of the Turkish campaign are exhibited.

try in Europe, Poland became more and more difficult to govern, especially as it was surrounded by enemies on all sides.

Poland's last king was Stanislaw Poniatowski who came to the throne in 1764 with the help of Empress Catherine the Great of Russia who had defeated the Turks and conquered the Ukraine. King Stanislaw was entirely under the influence of Russia but he did encourage scholars and artists from abroad to visit Poland. He finally went into exile where he died, a sad old man, in 1798.

The Polish people became restless because they feared Russia's growing influence, so they set up a resistance movement called the Confederation of Bar to oppose the Russian troops who had come to Poland. The Confederation eventually broke up and the first partition of Poland took place in 1772. Russia took the eastern provinces, the Austrian Empire took the southern parts of Poland, and Prussia annexed Pomerania. Shaken by these disasters, the Polish people introduced reforms in education and the army and

38

An equestrian statue of Josef Poniatowski, one of Napoleon's marshals.

voted, on May 3, 1791, for a new constitution. That day has since become Poland's National Holiday. Russia and Prussia disliked Poland's new constitution and so carried out the second partition of Poland, seizing new Polish provinces, in 1793.

Polish patriots planned an insurrection and chose as their commander General Tadeusz Kosciuszko who had served as an officer in the American Revolution. Kosciuszko led his army in April 1794 and defeated the Russian troops in the battle of Raclawice. But in the end Kosciuszko was defeated. Poland lost its freedom in 1795 and that year marked the date of its third partition.

The Polish nation never lost hope of regaining freedom. A Polish army under the command of General Jan Henryk Dabrowski was formed in Italy. One of the officers there, Josef Wybicki, wrote a song beginning with the words *Poland is not yet lost*. It was called at first *The Song of the Polish Legions in Italy* and later became known as Dabrowski's *Mazurka*. When Poland became independent many years later, in 1918, the song was adopted as Poland's national

39

anthem. It is still Poland's national anthem and this is stated in the Constitution. The first verse of the anthem is:

> *While we live she is existing*
> *Poland is not fallen;*
> *We will win with swords resisting,*
> *What the foe has stolen.*
> *March, march, Dabrowski,*
> *From Italy's plain,*
> *Our Brethren shall meet us*
> *In Poland again.*

At the beginning of the 19th century Napoleon started his military campaigns in Europe and wanted to re-establish the Polish state. General Dabrowski and his army joined Napoleon's forces and they arrived together in Poznan in 1806 where they were cheered by the town's population. After Napoleon's defeat at Waterloo, a peace conference or congress was held in Vienna in 1815. Unfortunately for Poland this decided on its fourth partition, with a large part of Polish territory going to Russia. A century was to pass until Poland regained its independence in 1918 after the First World War.

5

Death and Destruction

When the First World War broke out in 1914, the forces of the German kaiser were joined by the troops of the Austro-Hungarian Empire among whom were many Poles. Polish legions were formed to fight alongside the Austrian army. Much of the fighting was done on the eastern front against Russia but some Polish soldiers deserted rather than fight for their German and Austrian masters, especially as the Russian army also included Poles from the eastern provinces.

The downfall of the czar in Russia and the victory of the Bolsheviks in 1917 had a deep influence on the future of Poland. The new revolutionary Russia was the first to recognize Poland's right to be a free and independent country.

The end of the First World War meant the end of the German Empire and also of the Austro-Hungarian Empire. In October 1918, a Polish council was set up in Warsaw. The great task facing the Poles was to reunite the parts of their country which had belonged to the Germans, Austrians and Russians. The great

Polish leader Ignawcy Paderewski arrived in Warsaw and was asked to form a government, which lasted for the greater part of 1919. Jozef Pilsudski was elected head of state by the New Constituent Assembly or parliament. The new Polish Republic was also given a new constitution.

The Treaty of Versailles in 1919 officially gave Poland its independence though it took several more years to settle Poland's borders with its hostile neighbors. A war even broke out between Poland and Russia as no agreement could be reached on where the border should be. But finally peace was made. The new Poland had a population of over 30 million with a territory nearly the size of France. Access to the Baltic Sea was ensured by a narrow strip of land called the Polish Corridor but, unfortunately for the Poles, the city of Danzig (later to be known as Gdansk) at the mouth of the Vistula was turned into a free city under international control.

Misfortune came again to the Polish people in 1939 when Adolf Hitler's troops marched into Poland, because he demanded Danzig and the region of Pomerania. Britain and France then declared war on Germany and so the Second World War began. The ill-equipped Polish army was no match for German tanks and airplanes, and half the country, including Warsaw, was soon occupied. The eastern half of the country was invaded by the Russians. In June 1941, the Germans attacked Russia and so the whole of Poland fell under the Germans, who behaved cruelly toward the population. The Germans set up a reign of terror and appointed as Governor of Poland General Heinrich Frank who had his headquarters in Krakow.

The modern center of Warsaw rebuilt after the Second World War. The tall building is the Palace of Culture and Science.

The Germans built terrible concentration camps where thousands of prisoners perished in gas chambers. The most notorious of these camps were at Treblinka, Sobibor, Majdanek and Oswiecim where three million people lost their lives. People in Poland still recall how the Germans destroyed whole villages and marched the people off to concentration camps.

In spite of the terror, the Polish people continued the fight against the Germans, blowing up locomotives, trains, and trucks, and attacking German soldiers. The Polish partisans lived in forests and in the mountains and were supplied with arms dropped by British airplanes at night. Polish units fought the

43

Germans abroad in the Battles of Narvik in Norway, at Tobruk in the African desert, and in the battle for the abbey at Monte Cassino in Italy. Polish pilots took part in the Battle of Britain. There were also Polish soldiers fighting with the Russian army on the eastern front.

The Red Army in Russia, together with Polish units, eventually drove back the German invaders and, in 1944, they advanced across the old Polish border in pursuit of the Germans. In their retreat the Germans blew up bridges and railways and set fire to factories and houses. They destroyed crops and slaughtered cattle and horses. Warsaw was completely devastated and hardly a single building was left standing because of the fierce fighting between German soldiers and Polish partisans. Warsaw was liberated by the Russians on January 17, 1945.

The Second World War ended on May 9, 1945 when all the German armies capitulated. That day has since been observed as Victory Day. A conference was held at Potsdam (attended by Great Britain, the United States, France, and Russia) to decide the new borders. The conference decided that the lands east of the Odra and Nysa rivers should belong to Poland. Part of what used to be East Prussia became Polish territory but part of the eastern territories, including the important cities of Wilno and Lwow and part of the oilfields, was annexed by the Russians.

6

Poland Reborn

A new Polish state came into being after the Second World War. The first government of this state was called the Polish Committee of National Liberation. The communists now saw their chance in Poland. With the help of the Russian army stationed in the country, they took over the government and banned all other political parties. At first, they persecuted church leaders and also banned

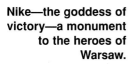

Nike—the goddess of victory—a monument to the heroes of Warsaw.

newspapers and books which printed criticism of their style of government.

A new constitution was approved by the Sejm on July 22, 1952. It laid down the rights and duties of Polish citizens and established the rules and system of government. The new constitution also proclaimed that Poland's coat of arms would continue to be a white eagle on a red background and that the country's flag would be white and red—divided horizontally, with the top half white and the lower half red.

After the fall of communism in 1989 there were many reforms in Poland. A new constitution was ratified in May 1997. The president is elected by popular vote for a term of five-years. The president appoints the prime minister and deputy prime minister, who then have to be approved by the Sejm.

There is a Council of Ministers to assist the president and prime minister, and they are also appointed by the president and approved by the Sejm.

The National Assembly (parliament) consists of the Sejm, which has 460 seats, and the Senate, which has 100 seats. The members of the Sejm are elected for four-year terms using a complex system of territory representation. The members of the senate are elected for four-year terms by a vote on a provincial basis.

The president appoints the judges to the Supreme Court. Other judges are appointed by the Sejm for terms of nine years.

The flag of Poland has remained the same since the changes in government have occured. Men and women are equal in modern Poland. Women are entitled to be given work on equal terms with men and to be paid the same wages. They have the same right as men to take part in public life.

Poland has social services which are for old people, the disabled, and the victims of natural disasters like floods and fire. In this, the state services are helped by the Catholic association *Caritas*.

Poland was a founding member of the United Nations Organization (U.N.) and has always taken an active part in the work of United Nations committees. Until 1989 Polish representatives on the Security Council or at the United Nations General Assembly voted, though sometimes reluctantly, for resolutions put forward by the Soviet Union or other communist or left-wing countries, with the exception of China. Modern Poland was regarded by the western countries as a supporter and close follower of the Soviet Union. Today Poland is a new member of the North Atlantic Treaty Organization (NATO) and votes independently in the United Nations.

7

Industry, Transportation, and Mining

In the early years of this century Poland was mainly an agricultural country. Most of its people were farmers. Since then it has become an important industrial country and today its goods are exported all over the world.

The Second World War from 1939 to 1945 brought ruin to industry just as it had begun to develop in the big cities. The border changes and war losses reduced Poland's population from 35 million in 1938 to 24 million in 1946, though since then the population has increased steadily.

The first task of the nation after the war was rebuilding towns and factories. Production was planned by experts in Warsaw. Since then there have been several economic plans in Poland with the aim of developing new industries and expanding production. The first big plan was the Three-Year Plan of Economic Rehabilitation which lasted from 1947 to 1949. That plan was followed by the Six-Year Plan of Economic Development and

A modern housing estate in Warsaw (the buildings in the foreground were reconstructed after the Second World War).

Building the Basis of Socialism which started in 1950 and ended in 1955. The Five-Year Plan of Economic Development followed the next year.

Branches of industry that were unknown before the war have been created in Poland. Machine tools, mining equipment, cars and trucks, tractors and ships, plastic materials, and artificial fibers are manufactured.

One of Poland's main industrial areas is the Silesian Dale and the city of Wroclaw, formerly known as Breslau. This region of coal and zinc mines, factories, and tall chimneys has become one of the great industrial centers of Europe. There are many iron-works and steelworks in Silesia, and factories producing electrical equipment, chemicals and textiles. Wroclaw has paper mills, flour

49

mills, and factories that make railway equipment and heavy machinery.

The Upper Vistula Basin forms another major industrial area. The city of Krakow with its industrial complex called Nowa Huta produce engineering goods. These goods vary from nuts and bolts to girders and turbines. There are linen mills and shoe factories in Krakow. There are also many food factories curing bacon and making the famous Krakow sausages.

Lodz, the second largest city after Warsaw, is an important center of the textile industry. There are cotton and woolen mills and factories producing artificial silk. Bialystok in the northeast of Poland also manufactures textiles, including blankets. Czestochowa, famous in the past for its monastery, now has jute mills.

Gdansk, Sopot and Gdynia on the Baltic coast form an industrial "trio" with their extensive docks, facilities for loading coal and iron ores, fishing companies and shipping firms. Gdansk was the cradle of the workers movement, Solidarity, and the first workers' strike occurred at the former Lenin Shipyards there in 1980. The Komuna Paryzka or "Paris Commune" shipyards, one of Europe's most modern shipyards, where oil tankers are built, are in Gdynia.

Szczecin (formerly Stettin) is Poland's largest port center at the mouth of the Odra River. Barges laden with coal and timber come down the Odra and deliver their coal at the long quays and piers. Nearby there are warehouses and shunting-yards and powerful cranes loading and unloading the cargoes of ships. Szczecin has

50

A view of Szczecin, one of the main Baltic ports, at the mouth of the Odra River.

boatyards which build fishing boats and yachts. The Odra shipyards make oceangoing vessels. There are also factories making machinery, chemicals, food products, and paper.

Poznan, in western Poland, is a center of the textile and leatherware industries. Further east is Warsaw, Poland's greatest industrial center, where all the factories have been rebuilt since 1945. Warsaw produces electrical machinery, and engineering goods of all kinds. Warsaw also has printing works, textile mills and factories processing food, drugs, and chemicals, not to mention the many gas and power plants. The land around Warsaw produces fruit and vegetables which are processed in the city's factories.

Lublin, in the east of Poland, is the industrial and cultural center for southeastern Poland.

Because the land is mostly flat, transportation has always been easy in Poland. As early as the 18th century two canals called Krolewski and Oginskiego were built, linking the Baltic and Black Seas via the Vistula and Dnepr rivers and their tributaries. Today Poland has 2,796 miles (4,500 kilometers) of waterways used by barges, tugs, and small steamers.

Between 1815 and 1863 the building of metal roads was started in the old kingdom of Poland. The roads built then are still in use today. But also today there are new east-west highways under construction and there are a few four-lane highways—though they are an exception. Driving conditions in Poland are not good.

Many of the small side roads have no hard surface and they

Lublin, an industrial center of modern times, has many beautiful historic buildings too. This picture shows the cathedral.

become deeply rutted after the heavy spring rains. In winter, after heavy snow falls, the only travel possible is by sled. For centuries people have traveled along the snowy roads and over frozen rivers in a sled drawn by a Konik pony. Round their necks, the ponies wear bells which tinkle merrily as they trot over the crisp and frozen snow.

The first railway line in Poland, the famous Warsaw-Vienna railway, was started in the years 1815 to 1848. Building the track was slow because the work was done by gangs of navvies and there were no modern machines or cranes to carry and lift the heavy rails. Today the total length of track is about 15,195 miles (24,313 kilometers). Powerful diesel engines are used on most lines. Warsaw, Wroclaw, Poznan, Katowice, and Krakow have extensive marshaling yards and railway workshops. Today, it is possible to travel by train from Warsaw to Moscow and from there to Siberia and the Far East, or to Berlin and Paris in the west. There are also direct trains to Prague, Vienna, Budapest, and Bucharest in the south.

Most of Polands major cities have an airport and they are connected to each other by internal air routes. The Polish airline LOT was founded in 1929. LOT means "flight" in Polish. LOT airlines run services connecting Warsaw with the Middle and Far East, the United States and Canada, Africa and most capital cities of Europe.

Poland used to be one of the big oil-producing countries in Europe. The oil industry in Poland was started far back in 1853. As there were no cars in those days, nobody was interested in dis-

A view of Katowice—a modern industrial center. To the right is the stadium and to the left a monument to the Silesian insurgents.

tilling gasoline out of oil. The oil extracted was used exclusively to make lamp oil or paraffin and lubricating oil for machinery, carts, and railway engines. The Polish oil fields are in the southeast, in the foothills of the Carpathian Mountains. The production of crude oil and natural gas today has dwindled because the deposits have been exhausted and are not enough to cover the country's needs. So extra oil and gas are imported through pipeline from Russia.

Hard coal (anthracite) continues to be Poland's most important mineral, though considerable quantities of brown coal (lignite) are also mined. The coal is mined in Silesia, in the Dabrowa and Krakow regions. The pits are seldom more than 1,000 feet (300 meters) deep which makes it easy for the miners. Some of the coal is exported to the Czech Republic, Slovakia, and Germany or shipped abroad by way of Szczecin and Gdansk.

Poland also has important iron quarries, and zinc and copper mines. Rock salt deposits are found in Silesia, at Wapno and near

Krakow. The oldest and richest rock-salt mine is at Wieliczka. Salt was mined there in 1040. The mine's underground passages total more than 62 miles (100 kilometers). The Wieliczka mine produces crystal salt, fit for human consumption, and rock salt, widely used for cattle and in industry.

Sulphur, lime, and chalk are also mined in Poland. In addition, marble is cut in blocks out of the mountainside in Silesia; while granite, used for making curb and cobble stones, and monuments, is quarried in the Tatra Mountains.

8

Farming and Village Life

As we have seen, not very long ago Poland was a farming country with over half its population employed on farms. Today, just one-fourth of the working population are employed in agriculture. Eighty-five percent of Poland's farming land is owned by peasants who do all their own work. There are some larger farms, but the large state-owned farms went into decline after 1989. There was a drought in the early 1990s that furhter crippled Poland's agricultural industry and Poland began to import a large portion of its food supply.

In some parts of the country the peasants have found it useful to join their narrow strips of land and pool their livestock and machinery into a farming cooperative. They can then make use of big tractors to plow and combine-harvesters to reap crops over a large area. They also jointly earn enough money to buy fertilizers, seeds and tools.

Traveling in summer through farming country not far from Krakow the visitor sees prosperous-looking farms and fields of

Prosperous-looking landscape in one of Poland's farming regions.

rye, oats, wheat, and potatoes. Rye is Poland's most popular crop because it can grow in poor soil and endure the cold weather. Rye flour is used to make the dark bread which is eaten all over Poland and even exported abroad. The long fine straw of the rye plant is used for thatching houses by skilled craftsmen who often make patterns along the ridges or above the eaves. Thatch keeps the house warm in the bitterly cold winters of the Polish plains; in the hot dusty summer it keeps the rooms cool.

Oats are grown in northeastern Poland while wheat mostly grows in the south. Potatoes are grown for food and for making vodka which is the most popular drink in Poland. The country also produces large quantities of wheat, sugar beets, hemp, and rape which is grown as food for sheep. Rape-seed also produces an oil which is used as a lubricant.

On some of Poland's small farms, the peasants still plow and harrow their fields with horses. There are over two million horses in Poland ranging from Arabs to the humble native ponies. There are proportionately more horses and ponies in Poland than in any other European country today and their value is going up because fuel for cars and lorries has become so expensive. Horses are also slaughtered for meat.

The Tarpan, the European wild horse, was a native of the Polish plains until it almost became extinct in the 1800s. However, two wild herds of this small gray-colored animal still survive in Bialowieza National Park and people say that a few tame Tarpans are owned by peasants in remote villages.

Two breeds of ponies, the Hucul and Konik, or Little Horse, can be found all over southern Poland. They are popular with small farmers because they are hard workers and economical feeders.

The Muraköz breed of horse is widely used in southern Poland. These animals have a thick neck and are sturdy farm horses. They were originally bred in Hungary. The Muraköz are mostly chestnut, bay or black. Further north, farmers use the large and handsome Wielkopolski breed which can be put in harness or ridden. These horses are good-natured and they mature early.

Hay and lucerne are valuable fodder crops which are fed to horses and cattle. Some Polish peasants still cut the hay with a scythe and turn it with a fork. When the hay is ready it is made into ricks and thatched, or put under cover in rickyards. Hay on larger farms is, of course, cut and baled by machinery.

In spring, when the trees are in blossom and the meadows studded with yellow and white flowers, the air is full of the buzzing sounds of bees. A beehive or two is a common sight in a Polish village backyard. There are nearly two and a half million hives in Poland today and the honey produced by them is sold locally or sent to the bee farms to be ripened and packed for export. Beeswax is used for making furniture polish and for candles used in churches.

Poland produces large quantities of fruit. The valleys of southern Poland have rich apple, pear, apricot, and plum orchards. Soft fruit, such as gooseberries, and red and black currants are grown by peasants on their farms. Most of the fruit is sent to market or to jam factories. Children gather fragrant wild strawberries in the mountains and small blue-black bilberries which grow on heaths and stony moors and in mountain woods.

Small cucumbers called gherkins, suitable for pickling, are very

A country inn built in the old style.

popular in Poland and are grown throughout the country. The gherkins are pickled in sour-sweet vinegar flavored with garlic, horseradish, and a herb called dill. Jars of pickled Polish gherkins are exported all over the world. Poles like to eat the pickled gherkins with sausages or to chop them into thin slices and make soup with them.

Now let us visit a village in the plains of eastern Poland. The village is called Krasnik and has a busy marketplace with a few shops, the post-office, a café, the House of Culture (where village dances are held) and the church. It is approached by a straight road with open, unfenced country on either side. Poplar trees line its sides and in the ditches there are nettles and thistles mixed with red poppies. The noise made by bumblebees and crickets fills the air. As it is Saturday, horse-drawn wagons and loaded carts trundle along the road on their way to the marketplace.

On the edge of the village there is a manor house. The drive leading up to the house is flanked by lime trees and the house has a balcony in front with creeping vines and a flower garden. At the back, there is a farmyard with stables and a hayloft, a barn for farm machinery and pig sties. Hens and geese roam freely in the yard.

Some houses of the village are made of brick, plastered and whitewashed, with a thatched roof, while others are made of logs that are closely fitted together. The smaller houses have only two rooms and a verandah. Inside the houses are wooden beds with pillows and eiderdowns in home-spun linen covers. There are also

A village family in traditional costume outside their wooden house.

wooden tables, chairs and chests. Colored plates hang on the wall; and, above the bed, there is a crucifix and a picture of the Holy Mother holding her Baby.

As it is market day, the village square is full of people and stalls. Some of the stalls sell a variety of vegetables, strings of onions, fruit and herbs. Other stalls sell brightly-colored toys carved from wood, whistles made from clay, mirrors, ashtrays and plates. Whips, saddles, boots, belts and other leather goods are on display at another stall. Pots and pans, wooden spoons and tools are laid out on a big canvas sheet. Pigs in pens squeal and hens in cages cackle while the market men and their customers clinch their deals.

All around, the cart horses wait patiently in the shadow of the trees away from the flies, munching hay and sugar beet tops. About noon, the market is over, everything is packed up and loaded into carts and the stalls are dismantled. Soon, only a few pieces of waste paper, odd broken plates and dried-up leaves show where the market has been. Some people stay behind. They drink beer in the café, or eat a meal of beetroot soup with sour cream followed by puff-pastry stuffed with spiced cabbage. A gray bus arrives and parks in the middle of the square. A group of musicians come out followed by several young people. They have come for the dance to be held later in the day at the House of Culture.

9

Forests and Fisheries

Great forests once covered Poland but many trees were cut for building houses and for firewood. The tree stumps were then burned to clear the land for farming. Yet 29 percent of the total area of the country is still covered with forests, some of which are primeval—belonging to the first age of the world. Pine forests are typical on the sandy soil of the lowlands. Some of the trees are 200

Logs which have been cut from Polish forests. They will be floated down to the saw mills from this collecting point.

years old. Pine trees produce a milky sap called resin. Cuts are made in the bark of the pines and the resin is collected in small cups placed underneath. Resin is used for making medicines, glues, and varnishes.

In parts of the country where the land is moist, leaf-bearing trees such as beech, oak and alder grow well. Ferns and bushes grow among the trees. The silver birch tree grows well in eastern Poland. There is one famous yew-tree called "Raciborski" reputed to be 1,000 years old. It grows with some ancient oak trees in the primeval Bialowieza Forest.

Coniferous trees, such as larch and fir, are found in many mountain regions together with the evergreen juniper shrubs. The juniper has prickly leaves and dark purplish berries. The oil from those berries is used for making medicines and in the manufacture of gin and varnish. Poland also has spruce trees which produce soft white wood. They thrive on wet clay.

Each year thousands of trees are cut by lumbermen wearing steel helmets. They do their work with chain saws. And each year thousands of young trees are planted to re-establish the forests. The trees are cut into cords—long logs—and pushed down the mountain side in a wooden or metal trough made slippery by running water. The logs travel down at terrific speed and land in a stream or lake below. Men waiting below lash the logs together to form a raft and steer them, with the help of huge poles that also act as rudders, as far as the sawmills.

Polish woodmen are great experts at making mountain trumpets. They take a long piece of dry spruce and hollow it out into a tube.

They bind it all around with the bark of a birch tree. When they blow the trumpet they can be heard for long distances; the sound is like thunder.

In summer, when the sun is hot and the grass and bushes are dry, there is great danger of fire. Whole forests have been set on fire and destroyed by careless campers. Forest rangers now forbid campers to light fires in forests. They also build watchtowers to watch for smoke. When they spot a fire they raise the alarm by informing the nearest firefighting station. When serious forest fires take place, troops may be called in with bulldozers and special equipment to put out the flames. Firefighting planes also spray

This fisherman has caught his fish with a rod and line but is using a net to pull it out of the water. Freshwater fishing is very popular in Poland.

chemicals which choke the flames. In this way many valuable forests are saved from destruction every year.

Fishing is an important industry in Poland. Fishermen go out in boats on lakes and rivers and catch mostly carp, perch, roach, and pike. Freshwater fishing takes place all over Poland, especially in the north where there are many lakes.

Deep-sea fishing is done by trawlers which go to the North Sea, the Atlantic Ocean, and even the Arctic Ocean north of Russia. They catch herring, haddock, and cod which are then processed on factory ships.

Gdynia is the headquarters of the Polish fishing fleet. It has large refrigerating plants, a Fish Research Institute, and several maritime schools.

Poland signed the Antarctica Treaty in Washington in December 1959. Since then, Polish scientists have taken part in exploring the icy seas around the continent of Antarctica, particularly for fish and whales. A few Polish whalers have been sent to fish in the Antarctic seas.

10

Warsaw and Krakow

Warsaw has been the capital of Poland since 1596. Excavations by archaeologists have shown that a castle stood in the Stare Brodno quarter in the 10th century and that in the 12th century there was a busy trading center on the right bank of the Vistula in the heart of Poland. At that time there was also a marketplace on the left bank.

The old city of Warsaw began to grow in importance from the beginning of the 14th century. So many new houses and shops were built that a century later the New City appeared outside the walls of the Old City. In 1413 the Masovian Prince Janusz moved his court from the Castle of Czersk to Warsaw.

The beautiful Wilanow Palace in Warsaw.

Warsaw was completely rebuilt after the end of the Second World War. This modern hotel is one of the buildings that rose from the ashes of war.

Merchants from eastern Europe and the valley of the Danube traveled through the Moravian Gap and the passes in the Carpathian Mountains into Poland and then down the Vistula River. They came in barges and galleys bringing grain and salt. At the same time, merchants from the Baltic came up the river bringing furs, fish, and amber in their boats. They all met in Warsaw to exchange their goods.

The most important event in the history of Warsaw occurred in 1596 when King Sigismund III transferred his capital from Krakow and took up residence with his whole court in the Castle of the Masovian Princes.

The 18th century was Warsaw's golden age. Poland's last king,

68

Stanislaw Poniatowski, did much to encourage master-builders and artists to come to Warsaw and build beautiful palaces. Industry also began to develop on a large scale. The first breweries, mills, and brick-kilns made their appearance at this time.

Unfortunately, Warsaw was damaged during Kosciuszko's armed insurrection in 1794 and the quarter of Praga on the right bank was burned down. Then, for the next ten years, the city belonged to Prussia. During the short period of the Duchy of Warsaw, the city was revived and expanded. For most of the 19th century Warsaw was ruled by the Russians. New industries were added and the population grew. There was much overcrowding and people lived in wooden houses. For three years in the First World War, from 1915 to 1918, Warsaw was occupied by the Germans. The Germans were defeated and at last Warsaw was free, the capital of an independent Polish state.

The greatest tragedy in the history of Warsaw took place in the Second World War. Warsaw fell to the German invaders in October 1939. From that day until their liberation by the Russians, on January 17, 1945, the people of Warsaw endured great suffering at the hands of the Germans. The sewers of the capital were used by Polish partisans as hideouts from which to fight the Germans. Later, the Jewish Ghetto was demolished and all its inhabitants killed. The city was utterly destroyed after the uprising in 1944.

Today, Warsaw is a city that has been completely rebuilt from its ruins. The houses and squares have been rebuilt as they were before and the bridges on the Vistula have been carefully reconstructed. A new big sports stadium was built in 1954. The white

The Krakow Cloth Hall.

building of the Palace of Culture and Science is the main landmark of the city with its massive wedding-cake architecture.

Warsaw is not only a place of great historical interest but also the center of the Polish government. The president has his residence in the Governor's Palace; the prime minister and other ministers have their offices in Warsaw. The Polish Sejm (parliament) holds its sessions in the Sejm House. The city's coat of arms is a mermaid with sword and shield.

Greater Warsaw, the modern city and its suburbs on both banks of the Vistula, has a population of over one and a half million. There is plenty of work in the factories and many young people come to Warsaw from all parts of Poland to study science, architecture, and many arts and crafts in the universities and the schools. Heavy traffic now flows through wide tree-lined boulevards bordered with busy shops and business houses. Trolley-buses and trams run in the streets.

Because it is so centrally placed on the banks of the Vistula, Warsaw is Poland's center for railway travel. The city is a good market in which farmers can sell their produce. In exchange, they buy cars and tractors, fertilizers, and all sorts of consumer goods.

Krakow, once capital of Poland, lies on the banks of the Upper Vistula in southern Poland. The city has always played an important role in the history of the country. Visitors from all over the world come to see it. They come to see ancient churches and palaces, museums, theaters and concert halls.

We know that primitive men lived in the Krakow area many thousands of years ago because their remains and remains of their tools have been found. These primitive men belonged to the Stone

St. Mary's Krakow.

Age which came after the Ice Age. Stone Age men used chipped stones as tools and weapons.

In about the ninth century Krakow was merely a cluster of wooden huts on a mound, protected by a deep moat and fence of wooden spikes. Then in 990 A.D., Krakow was taken over by King Mieszko I. A few years after the Poles had become Christians (in 966), it became the home of a bishop.

The Wawel, the old royal castle, stands on the steep hill overlooking the waters of the Vistula. According to an old legend, a fierce dragon lived under the walls of Wawel Castle and used to carry off the sheep grazing nearby. One day, the dragon was slain by a young Prince called Krak, and that is how the city got its name. Two earthen mounds called Krakus and Wanda lie on the outskirts of Krakow. People believe that the ancient King Krakus is buried there next to his daughter Princess Wanda who drowned herself rather than marry a German Prince.

During the 11th and 12th centuries the Tartars invaded Poland and laid waste to everything in their path. They plundered lands belonging to Krakow and laid siege to the city. During the siege, a Tartar arrow struck a Polish trumpeter at the top of the tower as he was sounding the alarm. The sound of the trumpet stopped as he fell dead. Ever since, by tradition, when the trumpet is sounded in Krakow, it suddenly stops in memory of the old trumpeter.

In 1306, Wladyslaw the Short became Prince of Krakow. In 1320 he finally became King and had himself crowned in Wawel Castle. From that time until 1596, Krakow became the capital of Poland and the city in which the kings were crowned.

King Casimir the Great established a university in Krakow in 1364. The great Polish astronomer Copernicus studied there. At that time, the city was becoming wealthy because merchants from eastern and western Europe met there to trade.

In the early 19th century Krakow was made a free city; this enabled it to modernize and lose its medieval character. Finally, Krakow was annexed by the Austrian Empire; it only returned to a free Poland after the First World War.

Krakow escaped destruction in the Second World War but many of its inhabitants were deported by the Germans and sent to concentration camps to be killed. Wawel Castle became the headquarters of the German governor of Poland. The city was liberated by Russian and Polish armies on January 18, 1945.

The Polish people regard Krakow as one of their most beautiful and historic cities. Its museums house valuable collections of paintings and sculptures as well as archives of Poland's history. Krakow Cathedral dates from the beginning of the 10th century and St. Leonard's Crypt houses the tombs of the Polish kings. Modern Krakow also has many schools for training doctors and scientists. Traditions dating from the Middle Ages are still observed today. The "Lajkonik," for instance, is a man dressed as a Tartar riding a dummy horse who dances through the streets and is cheered by the crowds.

Krakow is an important railway center and its market place has existed since the 13th century. It has flourishing industries and a population of 748,000 inhabitants.

11

Other Polish Cities

Many Polish cities and towns lie at crossroads or on important waterways and trading routes. They are famous for their castles which protected the citizens from swarms of marauding Tartars or the cruel Teutonic Knights. Let us visit a few of these centers.

Wroclaw (Breslau) is a large industrial city situated on the banks

The Town Hall of Wroclaw. Now a large industrial city, Wroclaw was originally named Breslau by the Prussians who ruled it for many years.

of the Odra in the middle of the Silesian Dale. It has a busy river port for barges and tugs, which unload coal, timber, and other bulky products.

The history of Wroclaw goes back a long way. It belonged to Poland in the late 10th century and was a stronghold in the Middle Ages. Later in its history Wroclaw was occupied by the Czechs, then the Austrians and finally the Prussians. On May 7, 1945 Wroclaw and Silesia again became Polish. The city had been badly damaged in the Second World War and many houses had to be rebuilt or repaired.

Modern Wroclaw is a mixture of old and new houses, streets and squares full of buses and traffic. The famous St. John the Baptist Cathedral was built in the 14th century, while the Town Hall dates from the 15th century. Wroclaw has a university, many schools, botanical gardens, and a zoo. Today Wroclaw has a population of 643,000.

North of Wroclaw, where the country is flat, is Poznan. The city is situated on the banks of the Warta River and was once the capital of Wielkopolska or Great Poland. In the 10th century Poznan was fortified and became the residence of a prince. In the old part

The impressive historic buildings of Szczecin seen from the river.

The cathedral of Szczecin, built in the 12th century. In the background is the modern part of this city which is now the largest in western Poland.

of the city there is a Gothic cathedral, the first church of its kind in Poland, which houses the tomb of King Mieszko I. The Town Hall is built in the Renaissance style with many stone and wood carvings. Unfortunately, the city was partly destroyed in the Second World War.

In spite of the damage, the city was rebuilt and has become prosperous again. It has several museums in which armor and weapons from the Middle Ages to the present day are on display, numerous scientific institutes, theaters, opera houses, and many churches. There are smart shops and cafes in the center. Today, Poznan is famous for the International Trade Fair held there each year. It has a population of 590,000. Its coat of arms is a gateway with three towers, surmounted by two saints and the white eagle. Traveling east from Poznan along one of Poland's main highways and then turning south, the visitor arrives at the city of Lodz–Poland's second largest city. Lodz is situated at the edge of the plains in the heart of Poland, where roads and railways meet. It was only a village in the 14th century, but it had become an important town and trading center by the 15th century.

At the end of the Second World War when Warsaw was a heap of ruins, the Polish government set up its offices in Lodz until Warsaw was rebuilt. The railway connecting Lodz with Warsaw is electrified. On the outskirts of the city there are cotton and woolen

The Town Hall of Gdansk

Buildings in the Old Town of Gdansk, reconstructed exactly as they originally stood.

mills producing cloth not only for the home market but also for the export trade. Lodz has a population of 834,000.

Now let us end our tour of Polish cities by visiting the Polish ports on the Baltic coast.

Szczecin is in the west on the Odra River, some 40 miles (65 kilometers) from the Baltic Sea. It is the largest city of western Poland and was badly damaged in the Second World War. The commercial port on the Odra and the shipbuilding yards were all rebuilt after the war.

Szczecin was already known in the ninth century. A Gothic cathedral was built there in the 12th century. In the 13th century many Germans came and settled there and they called the city Stettin. Today the city has a beautiful concert hall, several theaters, a museum, four universities, and botanical gardens. The city's coat of arms is a red eagle's head with a crown. There are many holiday resorts and sports centers for boating and yachting near Szczecin. The town has a population today of nearly 413,000.

East of Szczecin we come to the Gulf of Gdansk. The port of Gdansk, the largest Polish port on the Baltic, lies at the mouth of the Vistula some three miles (five kilometers) from the sea. The center of the town is between two small rivers, the Motlawa and Radunia.

Gdansk has been a busy trading town ever since the Middle Ages when it belonged to the Hanseatic League, a group of German ports which enjoyed privileges and freedom of trade. Ships from Finland, Sweden, Russia, England, and other countries came to the docks of Danzig, as it was then named, to unload their goods and pick up grain, timber, hides, coal, and iron. The history of Gdansk has been a stormy one. The Teutonic Knights captured the city for a time in 1308 when they slaughtered the Polish population. In 1526, the Polish king gave Gdansk special privileges and the city became famous for its furniture, clocks and jewelry.

Gdansk belonged to Prussia for a time, and then, in 1919, it became a Free City. Adolf Hitler used the difficulties created for the Germans in Gdansk as a pretext to invade Poland and so start the Second World War. Gdansk became Polish again in 1945.

The Old Town of Gdansk was completely rebuilt after the war. Today Gdansk has a medical school and a large public library. The Town Hall of Gdansk has an imposing entrance in its brick walls guarded by two stone lions which, according to legend, were carved by a stonemason named Daniel who wanted to remind the people of Gdansk that Poland was their mother and protectress. The city's coat of arms is two white crosses surmounted by a crown. Today Gdansk has a growing population of about 465,500.

To the northwest is the sister port of Gdynia which, until 1918, was a small village inhabited by fishermen. Today Gdynia is a bustling port with modern quays and cranes, grain elevators and shipbuilding yards. The population of Gdynia is over 251,500. The beaches near Gdynia are popular with vacationers. Because the ports of Gdansk and Gdynia do not freeze in winter they have become two of the most important shipping and trading centers in northeastern Europe.

12

Explorers and Scientists

Exploration and science have a long tradition in Poland. The University of Krakow gave the world great scholars who spread knowledge for the benefit of mankind. During the Middle Ages, in Poland and elsewhere, priests and monks in churches and monasteries taught children to read and write. By contrast, the children of princes, dukes and the nobility were taught to be gentlemen and knights. They thought that reading and writing was fit only for clerks.

A great Polish scholar Nikolay Kopernik—Copernicus—changed many strange ideas of his day about the world. He was born in the old city of Torun in 1473. His father was a rich merchant and his uncle a bishop. He studied law and medicine at Krakow University and privately he learned about astronomy. When he was 30, he visited Italy and studied in Bologna, Rome, and Padua. On his return to Poland, Kopernik wrote an interesting book about astronomy in which he claimed that the earth rotated on its axis and, together with the other planets, moved around the sun. Such ideas were hard to believe when people thought the earth was the center of the universe. Scholars and bishops poured scorn on Kopernik's theories. But his teachings were later accepted as true and now form the basis of modern astronomy which is an important science for the space age we live in.

To mark the 500th anniversary of Kopernik's birth, the Russians launched a satellite in 1973 which they called *Kopernik* and which circled the earth many times, sending back information about the weather and the upper layers of the atmosphere. The university in Torun is also named after him.

Sir Paul Strzelecki was perhaps the greatest Polish explorer, not only because he traveled all over the world but because he discovered and named places that had never been known before.

Paul Strzelecki was born in June 1797 at Gluszyna in the province of Poznan. His father was a landowner and his mother was a sister of the Archbishop of Gniezno. The young boy was sent to school in Warsaw where he stayed until the age of 16. He was restless at home and so decided to travel. In 1830 Strzelecki went to Scotland and explored the country. Scotland reminded him of the Polish highlands. He then went to England and set sail from Liverpool for the United States. He toured the United States and saw Niagara Falls. He went on to Canada and Mexico and then made his way by ship to Brazil, Argentina, and Peru. He crossed the Pacific to the Hawaiian Islands and arrived in New Zealand in 1839. In the same year, he traveled to Australia and found gold in New South Wales. Strzelecki toured Australia, measuring the height of hills and mountains and collecting geological

A statue of the astronomer Copernicus—Nikolay Kopernik.

specimens. He explored the Australian Alps and discovered their highest peak, which he called Mount Kosciuszko in honor of the great Polish patriot who fought for the freedom of his country.

Strzelecki also explored the greater part of Tasmania and the islands of Bass Strait. The highest peaks of Flinders Island were afterward named Strzelecki Peaks in his honor.

A Polish woman who achieved as great a name as Kopernik in the world of science was born in Warsaw in 1867. Her name was Manya Sklodowska and her father was a teacher of mathematics.

83

The young girl inherited a scientific mind from her father and at the age of 16 she was known as a scientific prodigy with a fabulous memory.

Manya Sklodowska went to Paris in 1891 to study at the famous Sorbonne University. She was very poor and she lived on bread and butter and tea. She met and worked with the French physicist Pierre Curie. They married and so she became Marie Curie.

Marie Curie had two daughters, Irene and Eve, who later helped her in her scientific work. Her husband died in 1906 but Marie Curie carried on the chemical research work which they had begun together. She discovered a new metallic element which was radioactive, and she called it polonium in honor of her native Poland. Later, she discovered radium. These discoveries led her to investigate dangerous radioactive materials, and it is thanks to her that X rays are now used in hospitals.

Marie Curie founded the Radium Institute in Warsaw and was twice awarded the Nobel prize. She died in 1934 of leukemia, a dreaded disease of the blood which she caught from exposure to radiation. The Polish authorities have named the university in Lublin the Marie Curie Sklodowska University in her memory.

Since her death, Polish scientists have continued research on radioactive isotopes used in medicine, industry, and farming. They have also given great priority to research on nuclear physics. Poland is a member of the International Atomic Energy Agency which is in Vienna.

Miroslaw Hermaszewski is a Polish explorer and scientist of more recent times. He was born at Lipniki in Poland in 1941. He went to the Deblin Air School and, qualifying as a pilot, he became a Group Commander in the Polish Air Force. In 1976 Hermaszewski was carefully examined and tested for his intelligence and health together with many other Polish candidates who wanted to become astronauts. He was chosen and sent to Russia to start training as an astronaut.

Then, on June 27, 1978, Hermaszewski and a Russian astronaut, Peter Klimuk, were launched into orbit around the earth in a spaceship called *Soyuz No. 30*. The Polish astronaut and his Russian colleague then docked with a Russian skylab circling the earth and spent seven days and nights in space. Hermaszewski had with him a copy of Kopernik's famous book *On the rotation of heavenly bodies*. When he was not working on board the skylab, he listened to the music of Chopin and other Polish composers.

The two astronauts then returned to their spaceship, and

The Chopin monument in Warsaw—a symbol of Polish patriotism.

returned to Earth, landing by parachute in a small capsule. Hermaszewski was taken to Warsaw where a huge crowd gave him a hero's welcome with flowers and flags. The Polish government awarded him a medal.

13

Churches and the First Polish Pope

Today, most of the Polish people are Roman Catholics. Christianity came to Poland in 966 when Mieszko I was king. He was married to Princess Dubravka who came from the kingdom of Bohemia and who was a Christian. The Bohemians became Christians themselves in the eighth century when missionaries came from Greece to preach to them. Through the influence of Princess Dubravka, King Mieszko was baptized with the whole Polish nation by missionaries from Bohemia.

Though the Polish people were baptized in the Early Christian Church, they adopted the forms of worship of the Church of Rome. In the 14th century Poland and Lithuania were united and many Lithuanians became Roman Catholics. But Lithuania included at that time Bielorussians and Ruthenians who were "Orthodox" or "true faith" believers. Some of these were united with the Church of Rome in the 16th century in the Union of Brest and came to be known as "Uniates." After the partitions of Poland, these Uniates were persecuted by the Russians and brought back to the Orthodox religion. Only those Uniates who lived under Austria remained in union with Rome.

It has not been easy for the Polish people to practice their Catholic religion. The Germans and Swedes who waged wars against the Poles had been Protestant since the 16th century and they persecuted the Catholics. In modern times, the Catholic Church was truly free in Poland for a short period from 1918 to 1939 when the Constitution declared that the Roman Catholic faith was the religion of the great majority of the nation and had a leading position in the state. An agreement called a Concordat was signed with the Holy See in Rome in 1925 and this guaranteed the freedom of the Catholic Church in Poland.

When the communist government first came to power in Poland, Roman Catholics were again persecuted. The Concordat with the Holy See was canceled, church publications were censored, and religious education forbidden. Archbishop Wyszynski of Warsaw, who was also Primate of the Roman Catholic Church in Poland, was put under house arrest for a time. Archbishop Wyszynski was a courageous and outspoken man who, in spite of many difficulties, made it possible for the Church to continue its work in the country.

In the course of centuries, Polish kings and princes built splendid abbeys and cathedrals, churches and monasteries. Masons and carpenters, architects and artists worked for many years to build these places of worship. Some churches were built in the Roman style with rounded arches and windows. Examples of this style are

A wooden church in southern Poland, built in traditional style.

the St. Mary Magdalen church in Wroclaw and the 12th-century church in Tum. Others, such as the cathedrals of Poznan and Szczecin, were built in the Gothic style. Artists were brought at vast cost from Italy, France, and Germany to decorate doors and pillars with elaborate carvings and embellish the tombs of kings and knights. The famous bronze door of the cathedral of Gniezno was cast at the beginning of the 12th century, while the altar piece of the Holy Virgin's Church in Krakow took 21 years to make, from 1477 to 1498.

The mountain districts of Poland are famous for their wooden churches. These churches have walls made of logs and their steep roofs go almost down to the ground. The roofs are covered in thin slats of wood called shingles, fitted neatly together like the scales

of a fish. The top of the roof is shaped like a dome and surmounted by a cross. Inside, the church is almost dark but visitors
can see tapestries hanging from the altar and precious icons on the
walls. In the middle of the church there are stands with brass or
other metal dishes on which the worshipers place lighted wax candles.

In the Middle Ages, monasteries and nunneries were built in the
hills and sheltered valleys of Poland, particularly around Krakow.
Some of the monasteries were fortified with towers and strong
walls to defend them against Tartar or Turkish marauders. At that
time, they were the only places where people could learn to read
and write. The monks and priests were scholars and craftsmen,
and people came to them for advice on how to weave cloth, carve
wood, or work with metals. Some of the monks were artists who
painted icons on glass or made beautiful furniture.

The fortified monastery of Czestochowa in the heart of Little
Poland is a place of pilgrimage. The monastery is situated on a hill
called Jasna Gora. It was besieged by the Swedes in 1655 but its
defenders, led by a humble priest, beat off the attackers.
Czestochowa is today the center of Polish religious traditions, and
the Icon of Black Mary inside the monastery, so called because the
face of the Virgin Mary is dark, is reputed to work miracles and
heal sick people. The monastery walls are covered with small silver or gold replicas of arms, legs, or other parts of the human
body. They were put there as gifts from crippled pilgrims who visited the monastery and were cured. Many pilgrims come to the

monastery to pray in front of the icon which still bears two small cuts made by the sword of a marauder in days gone by.

There are also shrines by the roadside for travelers. The shrines consist of a beautiful cross made of stone or wood. Sometimes the shrine is a pillar bearing the statue of the Virgin Mary or of St. Stanislas, Poland's patron saint.

Today, the Polish Roman Catholic Church has a great influence on the people. Old and young go regularly to mass. The churches are full to overflowing on Sundays and Saints' Days, and everybody participates in religious festivals when processions with banners and torches are held, followed by feasts. There are also Lutheran and Orthodox churches in Poland.

In 1978, dramatic events took place far away from Poland. A meeting of cardinals was held at the Sistine Chapel in Rome's Vatican City to elect a new pope, following the death of Pope John Paul I. The cardinals were locked in the chapel and supplied with food. They could not come out until they had made their choice. Every time a ballot was held they made a fire in a stove and the smoke coming out of the chimney was to tell the world what had happened—black meant nobody had been elected; white smoke meant a new pope had been chosen. After eight ballots white smoke was seen coming out of the chimney. Shortly afterward, the announcement was made that Archbishop Karol Wojtyla, a Pole, had been elected the new pope and would be known by the name John Paul II.

The news about the new pope was broadcast by the Vatican radio to the Polish people who could hardly believe their ears.

Karol Jozef Wojtyla was born on May 18, 1920 in the village of Wadowice some 75 miles (120 kilometers) from Krakow. His parents were poor people and his mother had died when he was nine. As a child, Karol attended the local secondary school (called a gymnasium) and was good at history, maths, Latin and Polish. He later went to Krakow University where he studied modern languages.

In 1941 Karol's father, who had been an officer in the Polish army, died. Young Karol was forced to work in a stone quarry and then in a chemical factory in Krakow, otherwise the Germans would have sent him to a labor camp. In the meantime, he had decided to become a priest and he went to live in the Archbishop's Palace in Krakow. After his ordination as a priest, he went to Rome and studied in the Angelicum University for two years. He returned to Krakow for more study and was awarded a degree in theology.

Karol Wojtyla then taught religion in Krakow and also at the Catholic University in Lublin. In his spare time he wrote poetry. In 1958, he became auxiliary bishop of Krakow and six years later he was promoted archbishop. Whenever he had time to spare he spent it hiking, cycling, or canoeing. He liked to live a simple life. By 1970, he was a cardinal of the Roman Catholic Church and took part in the meetings of bishops in Rome. In 1977 he inaugurated a new church in the industrial area of Nowa Huta before a crowd of some 50,000 people.

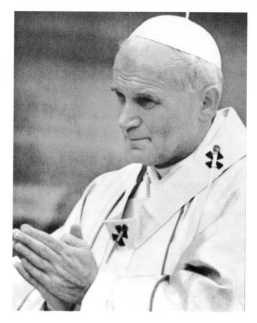

Pope John Paul II, the first Polish Pope.

When Cardinal Wojtyla was elected pope, the Polish political leaders sent him a telegram congratulating him on behalf of the Polish nation. The new pope soon became very popular. When he was crowned pope, the Polish head of state was present. The ceremony was shown on television in Poland. The streets and squares were quiet as the whole nation crowded around television sets and watched a former Polish factory worker being crowned pope and prince of the Vatican. It was the first time, for nearly 500 years, that a non-Italian had become pope. When a Turkish terrorist attempted to murder Pope John Paul II in Rome's St. Peter's Square in 1981, there was tremendous anxiety among the Polish people. But, after weeks in intensive care, the pope recovered. He was well enough, in June 1983, to visit his native Poland where he was cheered by huge crowds.

14

Sports, Arts, and Festivals

During their summer holidays, Polish boys and girls go camping by the lakes and in the mountains. Sometimes they stay in camps which have log huts, equipped with running water, kitchens, and electricity. But more often they set up tents in a forest clearing and cook their food by a campfire. At night, one of the campers keeps watch. In the silence of the night a broken twig may crackle and some wild animal may slip past. So nightwatch is regarded as an exciting adventure.

Tourists pony-trekking.

The campers are organized in patrols led by a scout master. When the patrol goes for a whole day's outing it usually has a special assignment. The patrol may have to find out from the local villages what changes have taken place in the past few years. On other occasions patrols go to places where there are commemorative plates and monuments in memory of the war dead. In these places, the patrols assemble for solemn roll calls of the dead and pay homage to those who fought and died for their country's freedom. The scoutmaster (who is usually a school games-master) also organizes boating, yachting, swimming, mountain climbing, and other outdoor activities.

Cycling is a popular sport in Poland because of the long straight roads. Cycle races are organized between various towns or in stadiums. Polish cyclists also take part in international cycling events.

Several Polish motorcyclists have become famous abroad. There are raceways for competing motorcyclists in Polish towns such as Poznan and Katowice and some of the best competitors have gone abroad to compete even as far away as Australia and New Zealand where motorcycle racing is popular.

Poland's cold and snowy winters mean that young people can go skiing and tobogganing in the mountains. The best ski slopes are in Silesia and at Zakopane at the foot of the Tatra Mountains where international competitions are usually held. Hotels have been built for the skiers at many mountain resorts and the slopes are equipped with ski lifts. People living in the cities can skate to music on open-air ice rinks which also have cafés.

There are other sports which are little known to young people in

A modern hotel in Zakopane.

Poland. Though the country has many ponies, there are few riding schools, or trekking centers except for tourists. There are only a few Polish tennis players because tennis courts can only be found in large cities. But table tennis is a popular sport.

Most Polish towns have a flying club where young people learn piloting, gliding, and parachuting. Poland is a suitable country for these clubs because of the many wide open fields without hedges or trees.

Today many people watch television or go to the theater or cinema for their entertainment. Modern music halls, dance clubs, and concert halls are popular in Poland because people are fond of music and proud of their traditional folk tunes. The Mazowsze song and dance company was known all over the world.

The National Philharmonic Orchestra in Warsaw and the State Philharmonic Orchestra and Choir in Krakow are also world famous. Smaller orchestras are found in Katowice and other Polish

96

cities. Musical festivals are held in Poland every year. One of the most famous of these is the International Song Contest held in Sopot on the Baltic Coast.

The Polish people observe many religious festivals and customs. At Easter, the women paint boiled eggs with traditional patterns of stars, circles, and crosses. On Easter Day, after the eggs have been blessed in church, people cut an egg into slices and give it to friends. Also on Easter day young men sprinkle girls with water according to an ancient pagan custom. On the second day of Easter it is the turn of the girls to sprinkle the young men with water.

Christmas is a great festival for the Polish people. It is traditional to eat fish and not meat on Christmas Eve. An old Christmas custom is to break wafers with relatives and friends. And everybody goes to Midnight Mass at Christmas.

A wedding in a Polish village is an occasion for much celebration. After the ceremony in church, the guests sit at long tables laden with roast duck or goose, pickled gherkins, loaves of rye bread, cheese, cake and fruit. There is always plenty of beer, wine, and vodka to drink. After the feast the musicians come in and the

Distinctive architecture in the Baltic resort town of Sopot.

A village wedding—an occasion for celebration—the traditional costumes worn are very elaborate and beautifully embroidered.

bridegroom and his bride start the dancing. They dance the traditional mazurka, krakowiak, and polonaise.

The Polish peasants love to wear traditional embroidered costumes and leather boots. The women continue traditional crafts; they weave carpets which they hang on the walls as a decoration and they embroider complicated patterns on their dresses. Decorations are made by cutting patterns out of colored paper. Young girls are good at this delicate work because they have good eyes and small hands. But young people living in the cities are the same as those in other parts of the world. They love to keep up with the latest fashions.

15

Poland and the Modern World

Poland became a communist country soon after the Russians defeated the Germans and drove them out in 1945. The country was ruled until 1989 by the powerful Polish United Workers Party which is just another name for the Communist Party.

Under the communist system the army and police were used to keep order and arrest any person who dared to criticize the government or Communist Party. Such people are called dissidents and either life was made unpleasant for them or they were imprisoned. Even trade-union leaders shared the same fate if they demanded higher wages for the workers of their union. From time to time, the workers held what have come to be called "bread riots" because food prices had gone up so much. Such riots took place in Warsaw and Gdansk in 1970 and 1976.

Strikes followed in 1980 and 1981. Lech Walesa, leader of the trade-union movement known as Solidarity and later to be awarded the Nobel Prize, demanded respect for union rights. The com-

A view of modern Poland—a street of new buildings in Warsaw.

munist government replied in January 1981 by disbanding all trade-unions. Later, on December 13th of the same year, martial law was imposed on the country and a Military Council of National Salvation was set up. There were violent clashes between miners, factory workers, and the army, and many people died. Lech Walesa was held under arrest for a time. Most papers, magazines and books published in Poland painted a rosy picture of the country and talked only of its achievements not its problems.

In foreign policy the Polish communist rulers sided with the Soviet Union and other communist countries such as East Germany, Hungary, or Bulgaria. Their representatives voted on

100

many occasions for Soviet Union resolutions at the United Nations Security Council or General Assembly and against Great Britain and the United States. The Polish representatives behaved in the same way at important talks on disarmament held in Vienna and Geneva. Poland, like a number of other communist countries, became an unwilling friend of the Soviet Union and so was called a satellite country by the western world.

But Solidarity did not give up its desire for democracy and independence from the Soviet Union. After the collapse of communism in 1989, Lech Walesa became president of Poland in the country's first free election since before World War II. The country changed its name back to the Republic of Poland, renamed streets, took down communist statues, and redesignated the use of Communist Party building—communism was purged from Poland.

The switch to a democratic society has not been an easy one but Poland took the lead for other eastern European countries. Poland has joined NATO and is reaching out to western countries for support and economic assistance.

Great improvements have been made in the standard of living in Poland. The country has made substantial progress in developing new branches of industry for the manufacture of scientific instruments, television sets, and computers. More crops are produced on farms and more houses are being built in towns and villages. There are free nurseries for children and free medical services for all. Poles can now own their own shops and so become richer by working hard. Many people in Poland today own brand-new cars, something which was impossible a few years ago. Some of the more fortunate Poles own cottages in the country or by the sea as well as town apartments.

A new constitution was adopted in May 1997 after 8 years of debate. Poland's economy is strong and the inflation rate is at an all-time low. The Polish people now face problems common to all western countries—increased crime, corruption, drug abuse, and the decrease of respect for religion. There are large gaps between the rich and the poor that need to be addressed also.

The Polish people are intensely patriotic and religious and have suffered many hardships in the past. They believe in being independent and they cherish their right to go to church freely and to decide their own affairs without foreign intervention. They now have the freedom to develop their towns and villages and industries in the way which will benefit the people most. And they will try to preserve their faith, their traditions, and their way of life in a rapidly changing world.

GLOSSARY

aurocks	An extinct large horned ox
Black Mary	The icon of Black Mary (a picture of a very dark blessed mother) is in a church in Czestochowa. It is thought to work miracles
Concordat	An agreement signed with the Vatican in Rome in 1925 guaranteeing the freedom of the Catholic Church in Poland
Copernicus	Born Nikolay Kopernik in Poland in 1473, Copernicus rewrote the books on astronomy with his theory of the earth revolving around the sun
Devil's Stone	A giant boulder measuring 82 feet across left in Poland by the Ice Age
hussars	Polish cavalrymen of the 15th century who wore steel armor and wings on their shoulders
icon	A religious image painted on a small wooden panel used as a devotional piece in Eastern European churches
marmots	Small animal that is a relative of the squirrel
moraines	Ridges of rock and earth left by the receding ice of the Ice Age

navvies	A British term for unskilled laborers
Polish Corridor	A narrow strip of land that connected Poland to the Baltic Sea
ricks	Stacks of hay left to dry in the open air
resin	A milky sap produced by pine trees that is used for making medicines, glues, and varnishes
Solidarity	Trade union movement lead by Lech Walesa, formed in 1980 to fight communist control of the government and industries of Poland

INDEX

107